W9-DEH-162

CHATTERING CHIPMUNKS

by Janet Piehl

Lerner Publications Company • Minneapolis

To Grandma Ele

Copyright © 2005 by Lerner Publications Company

All rights reserved. International copyright secured. No part of this book may be reproduced, stored in a retrieval system, or transmitted in any form or by any means—electronic, mechanical, photocopying, recording, or otherwise—without the prior written permission of Lerner Publications Company, except for the inclusion of brief quotations in an acknowledged review.

This book is available in two editions:
Library binding by Lerner Publications Company, a division of Lerner Publishing Group
Soft cover by First Avenue Editions, an imprint of Lerner Publishing Group
241 First Avenue North
Minneapolis, MN 55401 U.S.A.

Website address: www.lernerbooks.com

Words in *italic* type are explained in a glossary on page 30.

Library of Congress Cataloging-in-Publication Data

Piehl, Janet.
    Chattering chipmunks / by Janet Piehl.
       p.   cm. — (Pull ahead books)
    Includes index.
    ISBN: 0–8225–2420–1 (lib. bdg. : alk. paper)
    ISBN: 0–8225–2439–2 (pbk. : alk. paper)
    1. Chipmunks—Juvenile literature. I. Title.
    II. Series.
    QL737.R68P56 2005
    599.36'4—dc22   2004013947

Manufactured in the United States of America
1 2 3 4 5 6 — JR — 10 09 08 07 06 05

CHIP! CHIP! CHIP!

What's that sound?

It's a chipmunk. Have you seen
chipmunks in your backyard?

Chipmunks live in cities, parks, and yards. Some live in forests.

Chipmunks have furry tails.

They have sharp claws for climbing.

Chipmunks have stripes on their faces and backs.

Stripes help chipmunks hide
from enemies.

Can you find the chipmunk
on the tree?

CHIP! CHIP! CHUCK! CHUCK!

Chipmunks can make loud noises.
The noises help scare away enemies.

Animals such as weasels, foxes, bobcats, and snakes are chipmunks' enemies. This hawk is also a chipmunk enemy.

Chipmunks' enemies are called *predators*. Predators hunt and eat other animals.

Chipmunks eat fruit, nuts, and seeds. Sometimes they eat insects.

How do chipmunks find food?

Chipmunks find food on the ground and in bushes. They also climb trees to look for food.

Chipmunks use their long front teeth
to crack open seeds and nuts.
Then they eat them.

Sometimes a chipmunk keeps food in its mouth. It stores the food in *pouches* in its cheeks.

This chipmunk's cheek pouches are full of seeds and nuts.

What will it do with all of that food?

The chipmunk takes the food
home to its *burrow* below the ground.

The chipmunk saves the food
until winter.

Many chipmunks sleep in their burrows for most of the winter. They *hibernate*.

Sometimes they wake up and eat the food stored in the burrow.

A chipmunk was here!

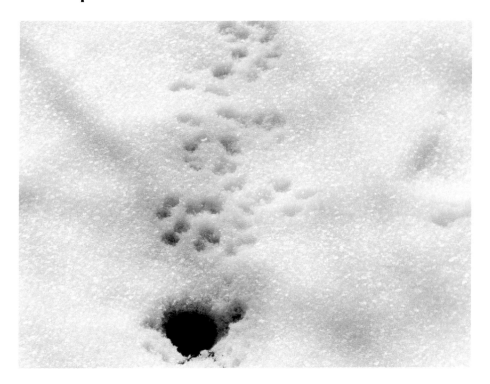

Some chipmunks go out on warm winter days. What do chipmunks do when the snow melts?

Mother chipmunks give birth to babies in the spring. At first, babies cannot hear or see. They stay in the burrow. The mother watches for danger.

The young chipmunks come out of the burrow after a few weeks.

The young chipmunks learn how to find food. They hide from predators.

The young chipmunks chase each other. They learn to climb.

Soon the young chipmunks are ready to leave their mother. They must find their own burrows.

CHUCK! CHUCK! CHUCK!

Maybe these chipmunks will make their new home in your yard.

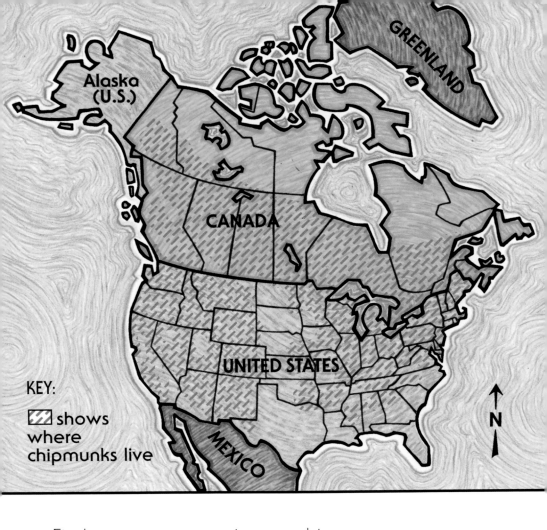

KEY:

⚡ shows where chipmunks live

Find your state or province on this map.
Do chipmunks live near you?

Parts of a Chipmunk's Body

tail

ear

fur

eye

nose

mouth

back
leg

cheek
pouch

claw

front
leg

paw

Glossary

burrow: place below ground where chipmunks sleep, raise young, and store food

hibernate: to sleep or be inactive during the winter

pouches: pockets on the inside of a chipmunk's cheeks. They are lined with fur. Chipmunks carry seeds and nuts there.

predators: animals that hunt and eat other animals

Hunt and Find

- a chipmunk **burrow** on pages 18–20
- a chipmunk **eating** on pages 13, 15, 27
- a chipmunk **sleeping** on page 19
- chipmunk **tracks** on page 21
- **young** chipmunks on pages 23–25

About the Author

Janet Piehl has lived in Delaware, Wisconsin, Minnesota, and France. Chipmunks also live in most of these places. She is also the author of *Formula One Race Cars*. Janet makes her home in Madison, Wisconsin.

Photo Acknowledgments

The photographs in this book are reproduced through the courtesy of: PhotoDisc Royalty Free by Getty Images, front cover, pp. 11, 12; © Gerry Lemmo, pp. 3, 8; © Gary H. Crabbe/Enlightened Images, p. 4; © Kim Taylor/Bruce Coleman, Inc., p. 5; © Len Rue Jr., pp. 6, 10; © Sylvester Allred/FUNDAMENTAL PHOTOGRAPHS, NEW YORK, p. 7; © Tom Vezo, p. 9; © Janet Horton, p. 13; © Joe McDonald/Visuals Unlimited, p. 14; © Steve Maslowski/Visuals Unlimited, p. 15; © Elinor Osborn, pp. 16, 18, 21; © Scott Nielsen/Bruce Coleman, Inc., p. 17; © KENT, BRECK P./Animals Animals, p. 19; © Dwight R. Kuhn, p. 20; © Wayne Lankinen/Bruce Coleman, Inc., p. 22; © Emma Ahart, p. 23; © John Shaw/Bruce Coleman, Inc., p. 24; © Eda Rogers, p. 25; © Richard P. Smith, p. 26; © Leonard Rue Enterprises, p. 27; © Gary Carter/Visuals Unlimited, p. 31